Heredity and Adaptation

Developed at
The Lawrence Hall of Science,
University of California, Berkeley
Published and distributed by
Delta Education,
a member of the School Specialty Family

© 2017 by The Regents of the University of California. All rights reserved. No part of this book may be reproduced or transmitted in any form or by any means, electronic or mechanical, including photocopying or recording, or by any information storage and retrieval system, without prior written permission.

1465675
978-1-62571-180-9
Printing 3 – 8/2017
Webcrafters, Madison, WI

Table of Contents

Readings

Investigation 1: The History of Life
Fossil Dating 3
An Interview with Jennifer Clack 11

Investigation 2: Heredity
Tree Thinking 17
Understanding Heredity 22
Mendel and Punnett Squares 28
Mapping the Human Genome 36

Investigation 3: Evolution
Adaptation 41
Natural Selection 53
What Makes a Scientific Theory? 60

Images and Data 69

References
Science Safety Rules 89
Glossary 91
Index 94

Fossil Dating

Earth is a very old place. According to recent evidence, it is around 4.6 billion years old.

Geologists call this vast amount of time **geologic time**. We know about past life by looking at remnants, marks, or traces of life-forms trapped in rock. Rocks hold the key to much of what we know about Earth's history and the **evolution** of life on this planet.

Fossils are evidence of ancient life. Most fossils are body fossils or trace fossils. **Body fossils** are rocks made from parts of the **organism**. The hard parts, such as shells, bones, and teeth, are most likely to become fossils. **Trace fossils** preserve evidence of activities. These rocks most often show tracks, tail-drag marks, burrows, impressions, and droppings.

Trace fossils, like these trilobites, are impressions left in mud that hardened to rock. Millions of years later, these impressions provide clues about how and where the animals lived.

Investigation 1: The History of Life

Colorful layers of sandstone, shale, and limestone make up the Grand Canyon. These sedimentary rocks contain evidence about how Earth's organisms and environments have changed over time.

Relative Dating

Most fossils are in **sedimentary rock**. Sedimentary rock forms over thousands, sometimes millions, of years. First, **particles** of gravel, sand, and mud settle on the bottoms of rivers, lakes, and the ocean. Little by little, those **sediments** pile up. Sediments also form in the desert. There, particles of dry sand drift and build up. Sediments sometimes cover dead organisms. The soft parts of organisms usually decay or are eaten. But the hard parts of organisms can remain. As sediments build up, the remains are buried deeper and deeper. The sediments become compressed and form rock around what remains of the organism.

Think Question

Fossils are relatively rare. Why don't we find more?

In the 1600s, Nicolaus Steno (1638–1686) figured out how sedimentary rocks form. The Danish scientist said that lower layers of rock were laid down first. They are usually older than the layers of rock above. This rule is the **principle of superposition**.

This principle allows **paleontologists** to compare fossil ages. A fossil in a lower layer of sedimentary rock is most likely older than a fossil found above it. This useful rule tells us the relative age of fossils. But it does not tell us how old a fossil or rock is. To do that, we look at the **atoms** in the rock.

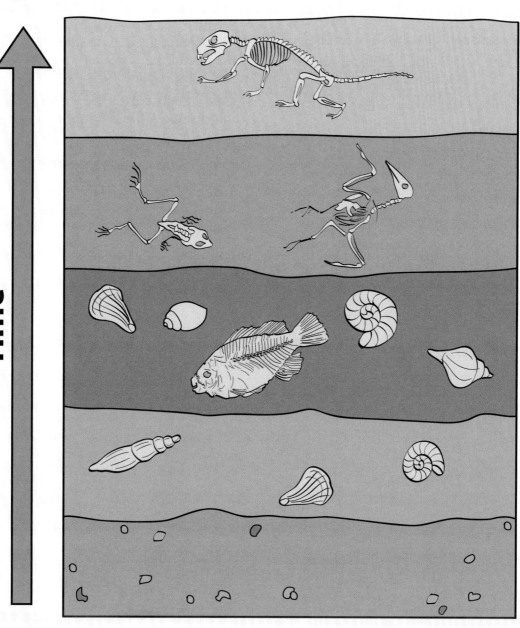

Relative dating compares the relative age of rock layers. In general, the lower the rock layer, the older it is and the older the fossils it contains.

Investigation 1: *The History of Life*

Absolute Dating

All matter is made of particles called atoms. Each atom is an element, such as carbon, potassium, uranium, or oxygen. Atoms are made of even smaller particles. In any one element, the number of smaller particles can vary. For example, carbon comes in three forms: carbon-12, carbon-13, and carbon-14. Each form has a different number of particles and is called an **isotope**. A **radioactive isotope** loses small particles at a fairly constant rate. Gradually, it changes into a different isotope. If it loses enough particles, it can change into a different element. This is radioactive decay.

Different radioactive isotopes decay at different rates. Depending on the isotope, it can take minutes to millions of years for half of the radioactive isotope in a sample to decay. This time is the half-life of the isotope. Isotopes with a long half-life are used to date rocks. For example, the half-life of uranium-235 is 704 million years. It takes

Scientists unearth fragile fossils with great care. Then they date the fossils by finding out the age of the surrounding rock.

704 million years for half of a sample of uranium-235 to turn into another element, lead. By measuring the ratio of uranium to lead in a rock sample, geologists can calculate when the rock formed. For example, if a rock is 50 percent uranium-235 and 50 percent lead, we know it is 704 million years old. Once you know the age of the rock, you know the age of the fossils in that rock.

Radioactive Decay

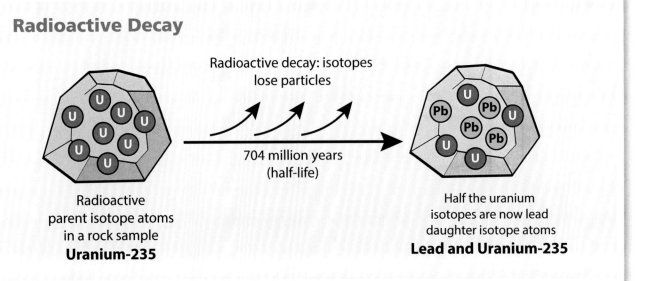

How old are fossils found in a sample of rock that is half uranium-235 and half lead? Would rock that had a higher percentage of uranium-235 be older or younger?

Could scientists use carbon dating to find out when this fossilized fish lived and died? It depends upon when it died. Carbon dating cannot be used to date fossils more than 50,000 years old.

Using Carbon-14

Carbon-14 has a half-life of 5,730 years. Paleontologists use this isotope to measure the age of fossils of organisms that lived in the last 50,000 years or so. Carbon-14 occurs in the upper atmosphere. All living organisms take in carbon-14. When an organism dies, it no longer takes in carbon-14. Instead, the carbon-14 in the remains decays to form another isotope of carbon, carbon-12. By measuring the amount of carbon-14 and carbon-12 in a sample of bone, hair, or wood in a fossil organism, scientists can determine how long ago it died.

Think Question

Why isn't carbon-14 used to date rock that is millions of years old?

Investigation 1: The History of Life

Geologic Time Scale

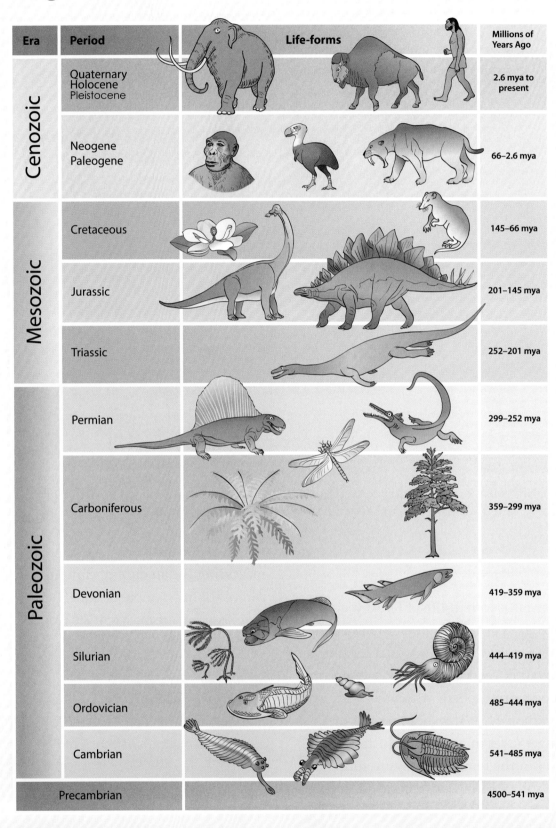

The geologic time scale is a model based on major changes in the fossil record.

Geologic Time Scale

Geologists and paleontologists use relative and absolute dating to estimate the ages of sedimentary rock and fossils. They have used the **fossil record** to develop a model called the geologic time scale. This model describes what we know about how old something is or when something happened during the history of life on Earth.

Study the geologic time scale presented on the left. It shows when different kinds of organisms first appeared. **Eras** are broad spans based on typical life-forms at that time. *Zoic* refers to animal life.

Crinoid fossils are some of the oldest fossils on Earth. These marine animals first appeared in the Ordovician period. Their descendants still exist in deep ocean waters.

Investigation 1: *The History of Life*

Ammonites are among the most common fossils found today. These squidlike predators lived inside spiral shells and were abundant in Earth's ocean from the Devonian through the Cretaceous period.

- *Paleozoic* means "ancient animal life." The Paleozoic era included a group of extinct marine arthropods called **trilobites**. This era included corals, **brachiopods** (an early marine organism with two hard shells or "valves"), early fish, and early amphibians. Plants of this era included early conifers and ferns.
- *Mesozoic* means "middle animal life." It is sometimes called the age of the dinosaurs.
- *Cenozoic* means "recent animal life." It is sometimes called the age of mammals because so many new kinds of mammals appeared during this era.

Each era is divided up into periods that reflect the fossil record. The geologic time scale is constantly being tested and revised. Geologists improve the model with new information from fossils and data.

Think Questions

1. The Devonian period is considered the Age of Fishes. Would you expect to find a tetrapod (animal with four legs) fossil closer to the Carboniferous period or the Silurian period? Why?
2. Certain kinds of shells are always found in rock layers lower than layers with plant fossils. Dinosaur fossils show up only in layers above or with plant fossils. What does this tell you about when these organisms lived?
3. What information would you need to find a more precise age of dinosaurs and plants?
4. Would you expect to find fish fossils in rocks 550 million years old? Explain.

An Interview with Jennifer Clack

The Devonian period in the Paleozoic era was the Age of Fishes. Several million years later, the Mesozoic era was the Age of Dinosaurs. How did vertebrates make the transition to land?

The fossil record provides evidence that life on Earth has gone through many breakthrough changes over the past 4 billion years. One of the most remarkable transitions gave rise to **tetrapods**, vertebrate animals with four limbs. The change was from a fish body plan adapted for breathing and navigating in water to a body plan like that of a mammal, reptile, or amphibian that provides for movement on land. How did vertebrates colonize the land? What are the transitional forms that tell us the story? Who were the aquatic ancestors of the tetrapods? Many paleontologists have studied this process and searched for evidence of the details of this transition. It is still an active area for research and discussion.

How did vertebrate life-forms move from aquatic to terrestrial environments millions of years ago?

Investigation 1: *The History of Life*

Jennifer Clack is a paleontologist who hunts for early tetrapod fossils. Her work has taken her all around the world, including Greenland and Scotland. FOSS caught up with Dr. Clack in December 2014.

FOSS: What made you become a paleontologist?

JC: It's kind of hard to say what first sparked my interest in paleontology. It was such a long time ago. But I do remember that when I was 9 or 10 years old, I was given a copy of Arthur Mee's *Children's Encyclopaedia* from the 1950s. It had chapters about the Paleozoic vertebrates. There were things that were so old and so bizarre.

FOSS: So it was actually a book that sparked your interest?

JC: I certainly was always interested in natural history. I used to watch science programs on the television. I would always read books on geology or natural history rather than novels.

FOSS: How did you end up looking for *Acanthostega gunnari*? I read that the fossils were first analyzed in the 1930s, but were lost in a storage area. You "rediscovered" the fossils.

JC: It's a little more complicated than that. There were two main groups of tetrapods from the Devonian period found in east Greenland. The one that has been known the longest is *Ichthyostega*. It was found in the late 1920s and studied in the early 1930s. It was described again by a Swede (Erik Jarvik [1907-1998]) in the 1950s.

Ichthyostega, the first tetrapod discovered and described, is a key transition between water and land vertebrates.

As part of her work at Cambridge University, Clack plans and prepares the vertebrate fossil displays at the zoology museum.

In his paper about *Ichthyostega*, Jarvik also mentioned a second group, known from just a couple of skulls or partial skulls. Those he named *Acanthostega*.

Now, nobody was able to study *Ichthyostega* while Jarvik was working on it. Paleontology has a sort of protocol. If you have the specimens and you're working on them, other people can look at them, but they cannot publish what they find. So *Ichthyostega* remained in a sort of limbo for decades because he didn't do anything more with it.

Now, as it turned out, when I came to Cambridge [in 1985], I was looking for something to study after I finished my PhD thesis. Luckily, it turned out that people across the road from me in the Earth Sciences Department had a drawer full of tetrapod fossils from a different expedition to Greenland. And among those were some specimens of *Acanthostega*. That gave me an "in" to the whole Devonian tetrapod question. Because it wasn't *Ichthyostega*, it was basically up for grabs. So I grabbed it!

In 1987, I managed to get together an expedition to Greenland to collect some more *Acanthostega* material.

Scientists all over the world have collections of fossils that can provide information to help us understand change over time.

FOSS: How did people learn that there were tetrapod fossils in Greenland?

JC: Well, it's quite a long story. It started in the 1800s. Both Denmark and Norway wanted to claim Greenland. They were particularly interested in mining minerals, so they sent lots of expeditions. Some of these early expeditions—particularly in the 1920s—happened to find fossils. Lots of them were fish fossils, but they finally recognized that tetrapods were there, too.

FOSS: Have any tetrapod fossils been found in other Devonian rock layers anywhere else in the world?

JC: Devonian tetrapods have now been found pretty much throughout the world. There are some in the United States, China, Australia, Russia, and the Baltics. So yes, they're all over the place. Some of the *Acanthostega* material we brought back from Greenland alerted people to some of the features they needed to look for. They actually discovered that they already had some in their museum drawers. The fossils just were not recognized!

Scientists think that during the Devonian period Greenland was united to North America and Europe in a single super-size landmass.

Many tetrapod fossils are found in Greenland, the world's largest island. It has an Arctic climate today, but when tetrapods were hauling themselves through its coastal shallows, it was swampy and much warmer.

Acanthostega gunnari's paddle-like limbs had no wrists or ankles. Their limbs were not weight bearing and wouldn't have allowed walking. Its broad tail shows that it was primarily a swimming animal.

FOSS: Is it your claim that the change in the skeleton of some tetrapods living in the water (fins to flipper-like structures with digits) happened before the transition from water to land?

JC: Yes. That idea is based on the fossil material as we found it. And, of course, it may be overturned by new fossil discoveries any day. But it makes sense.

There is only one possible challenge right now. Some people found in 2010 what they believe to be tetrapod trace fossils, called trackways, that are about 18 million years older than the body fossils we have found. We don't have any body fossils to go with these trackways, so we don't know what sort of an animal actually made them. It might have been another kind of animal using tetrapod-like locomotion. The animal appears to have been supported by water, however. And we don't know whether this animal would have been related to the things in the later fossil record, or whether it was a completely parallel branch. Until we get the body fossils, we won't be able to solve that one.

FOSS: So the question still remains! What about survival advantages of digits for an aquatic organism?

JC: Of course, we don't know precisely, but we can think of lots of things that would have made digits an advantage. First of all, you can grasp things. So if you're an ambush predator lurking in the weeds, you can hold on to the substrate (rocks, mud, or plants at the bottom of a body of water) and hold your station without moving. A fish with fins would have to keep its fins moving to stay still. So that's one possibility.

Also, digits allow organisms to spread the load of their weight on the substrate, especially if they are lifting their heads out of the water to breathe air. I think really those are the main advantages—spreading the load, and grasping vegetation, station holding, that sort of thing. You can go places where fishes find it difficult to go. So you can push your way through swampy waters, which are full of many kinds of plants.

Investigation 1: The History of Life

Svend-Erik Bendix Almgreen, Jennifer Clack, and Per Ahlberg in East Greenland. They found *Acanthostega* fossils at this site.

FOSS: People often think that scientists work alone, going by themselves to Greenland to dig out fossils in rocks. But what you're doing seems extremely collaborative. How do you collaborate?

JC: This has changed over the last 20 or 30 years. When I first started it was, more or less, people working by themselves. But with the growth of communication technology, it is so much easier to collaborate. And you become so much more productive. Really, collaboration is the key to progress these days.

FOSS: Do you think this is an exciting time in history to be a paleontologist?

JC: Well, it is! And we think we're really onto a project that will run and run. We hope that future generations will pick it up and run with it. But the focus seems to be dinosaurs in most people's minds. Very few scientists are talking about material from the time period we're looking at. So we'll see how it goes.

FOSS: We hope to provide you with some future scientists.

JC: That would be great, yes!

Tree Thinking

Humans love to sort things into groups. We sort foods into fruits, vegetables, meat, fish, and grains.

We group the clothes in our closets and then sort them by color. We sort vertebrates into mammals, birds, reptiles, amphibians, and fish. We sort sea shells found on a beach. For a long time, scientists have looked for systems for naming and grouping the diversity of life.

In the mid-1700s, a Swedish scientist named Carolus Linnaeus (1707–1778) organized living things into groups. He studied the **characteristics** of organisms. Then he grouped them, based on their similarities and differences. Other scientists began to look at fossils to study life's history. These scientists also used similarities and differences to classify life. As more organisms were found, scientists began to look for new ways to group them.

We group shells by their shape, size, and color.

A Branching Tree

About 100 years later, Charles Darwin (1809–1882) suggested organizing life in a tree pattern. The branching tree he sketched suggests that all life on Earth is **related**. Each branching point on Darwin's tree represents a **common ancestor** that divided into **descendants**. The root of the tree is the ancestor of all the **species** on the tree of life.

Today, scientists still use Darwin's "tree thinking" to model evolution. They make hypotheses about the relationships among species using a tree diagram called a **cladogram**. A cladogram maps patterns of relatedness based on patterns of shared **inherited characteristics** called **traits**. To be useful in making a cladogram, a trait must have been inherited by two or more of the species on the tree. It must have appeared for the first time in those species' **most recent common ancestor**. Extinct species can be included in the cladogram. Look at the portion of the example from class at the top of the next page.

In Linnaeus's time, scientists used similar observed traits to group organisms. So they might group the shark and the tuna. Both have eyes and fins. The cladogram, however, puts the tuna and frog together. This is because their most recent common ancestor had a bony skeleton. This trait (bony skeleton) first appeared in the most recent common ancestor of the tuna and frog. Sharks do not have bony skeletons. This tells us that the tuna and frog are more closely related than either is to the shark. We have to go farther back in time to find a most recent common ancestor for the shark, tuna, and frog. As you can see on the cladogram, that ancestor had vertebrae.

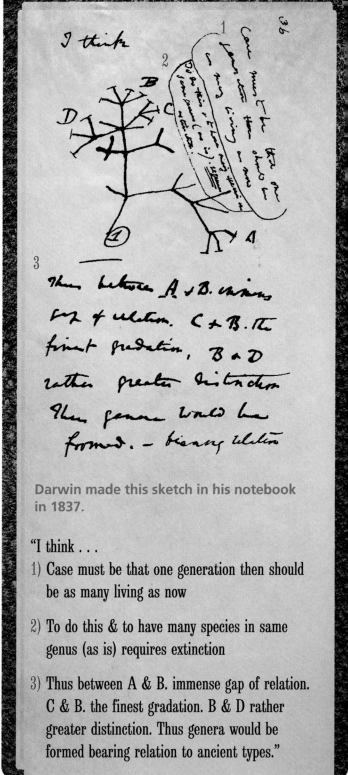

Darwin made this sketch in his notebook in 1837.

"I think . . .
1) Case must be that one generation then should be as many living as now

2) To do this & to have many species in same genus (as is) requires extinction

3) Thus between A & B. immense gap of relation. C & B. the finest gradation. B & D rather greater distinction. Thus genera would be formed bearing relation to ancient types."

Cladogram: Shark, Tuna, Frog

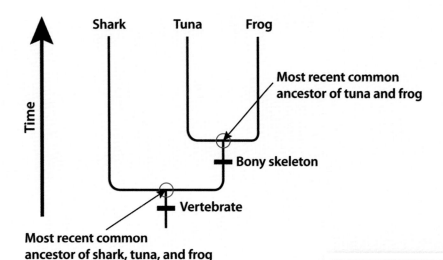

A cladogram is a model that illustrates the hypothetical relationship between common ancestors and their descendants.

What a Cladogram Tells Us

A cladogram shows the order in which traits appeared. It does not represent a simple line from ancestor to descendant. This is important to think about. The shark is not the ancestor of the tuna and frog. Instead, the shark, tuna, and frog share a common ancestor, at a branching point. Also, a frog is not "more evolved" than a shark. Each has different traits that allow it to thrive in its environment. Finally, a cladogram shows that one trait **evolved** in a group of organisms before or after another trait. For instance, vertebrae evolved before a bony skeleton did. To find out how long before, you need to look at the fossil record. The cladogram does not tell you.

A cladogram is only as accurate as the information used to construct it. Scientists may use physical characteristics, such as the arrangement of bones in a limb. Or they might use the kind of tissues or reproductive structures of a plant. Using only physical characteristics requires careful study.

A shark lacks the bony skeleton of a tuna or frog. That physical trait developed later in time than vertebrae, which appear in all three organisms.

What skeletal trait might separate the tuna from the frog? Does the tuna have limbs?

Both the frog and tuna have bony skeletons, but only one has lungs and legs.

Investigation 2: *Heredity*

Over the past 40 million years, the hyrax, manatee, and elephant have undergone dramatic changes from their common ancestor. Today they are adapted to life in vastly different environments.

A Hypothesis

Consider the hyrax, the manatee, and the elephant. At first, they don't appear to be related. But the hyrax and the manatee are two of the closest living relatives of the elephant. They share a most recent common ancestor. What evidence supports this relationship? Careful observation of physical traits reveals several unusual shared characteristics, such as small tusks, similar toenails and bone shapes, and excellent hearing. **DNA (deoxyribonucleic acid)** analysis shows similarities among the hyrax, the manatee, and the elephant. In addition, some 40-million-year-old fossils in Africa provide evidence that a mammal about the size of a pig is the most recent common ancestor of these three species.

The fully aquatic manatee, or sea cow, is descended from a land-based, wading mammal.

You might think that an elephant's closest relatives are hippos or rhinos. You'd be wrong.

Cladogram: Hyrax, Manatee, Elephant

```
Hyrax              Manatee            Elephant
(Hyracoidea)       (Sirenia)          (Proboscidea)
```

A cladogram is likely to show surprising relationships among organisms.

A cladogram is a hypothesis about the relationships among groups of organisms. It represents our best model of evolutionary relationships. It changes as new fossils or genetic information is discovered.

What cladograms reveal is that all living things are part of one huge bushy tree. They share a 3.8-billion-year-old common aquatic ancestor. As we gather more evidence and develop new methods for examining traits, our view of the tree of life becomes clearer.

> **Think Question**
>
> Look at the cladogram above. Place your finger on the intersection that represents the most recent common ancestor of the hyrax, the manatee, and the elephant.

Understanding Heredity

How is it that offspring resemble their parents? What is the mechanism that passes on characteristics from one generation to the next generation?

Every living thing looks pretty much like its parents. You look similar to, but not *exactly* like, your biological parents. Tomato plants look like their tomato-plant parents. Seems simple, but how does it happen?

Lambs resemble their parents because of heredity, the passing of traits from parents to offspring.

Clearly some kind of information passes from parents to offspring. This information includes instructions for how to develop into an adult like Mom and Dad. One **generation** passes information to the next. This **heredity** makes the new generation resemble the other members of their species, especially their parents. For thousands of years, people have recognized the usefulness of **inheritance**. They observed **variation** in the traits of organisms of the same kind. They selectively bred plants and animals so their offspring would have valuable traits. For example, a sheep with very long soft wool, or an apple tree with tasty fruit, would be selected for breeding. Often the offspring inherited the desired traits. But no one understood the actual mechanism that drives inheritance.

A tomato plant bears only tomatoes, but not all tomatoes are the same. Fruits with exceptional flavor, firmness, color, or ripening speed may become breeding stock.

Mendel's Research

Gregor Mendel (1822–1884) was born in a poor farming community in what is now the Czech Republic. He was a bright student, but his family did not have the money to send him to a university. In order to continue his studies, Mendel joined a monastery, where he studied the patterns of inheritance in pea plants. He conducted careful experiments over many years. He concluded that parent plants have a "factor" that transfers traits to their offspring.

Mendel's findings were ignored for many years. Scientists did not understand what made inheritance possible. Before long, however, advances in technology provided a new way to examine inheritance.

Mendel's pioneering work led to our understanding of heredity.

Mendel experimented with pea plants in the monastery garden for 8 years, from 1856 to 1863. He wanted to find out how traits such as flower color, pod shape, and stem length were passed from parent to offspring.

Egg and Sperm Formation

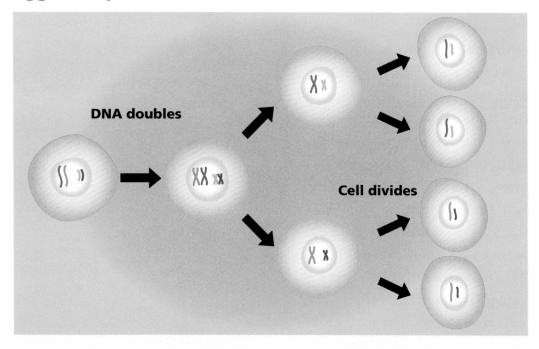

Sex cells are formed by a two-part cell division called meiosis. Duplicated genetic material in the parent cell is distributed among four daughter cells.

Discovery of Chromosomes

By the late 1880s, scientists were able to observe the nucleus of a cell. They discovered structures that looked like Xs and hot dogs inside the nucleus. They named these structures **chromosomes**. Scientists observed that when a cell prepared to divide, the chromosomes copied themselves. They separated into two identical sets of chromosomes. Each new cell enclosed a complete and identical set.

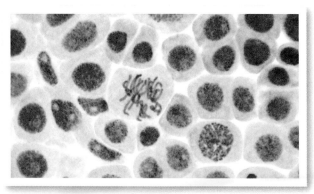

Chromosomes double before cells divide. These cells are typical body cells.

In 1902, Walter S. Sutton (1877–1916) observed that the chromosomes in a nucleus could be sorted into pairs. When some cells divided, the chromosome pairs separated and went into separate cells. These new cells were sex cells (egg and sperm). Each sex cell had just one set of chromosomes, not two.

This remarkable discovery confirmed Mendel's work. Sutton realized that Mendel's factors are located on chromosomes. During sexual reproduction, the chromosomes pass to the next generation, one set from the sperm, and one set from the egg. The fertilized egg has two sets of chromosomes, but the pairs are not exactly the same as those of either parent. The new pair of chromosomes is a mixture of the chromosomes of the parents. The information for an organism's traits is on the chromosomes it receives from both its parents.

A Remarkable Molecule

How do chromosomes carry information? The answer lies in a remarkable molecule, deoxyribonucleic acid (DNA). DNA was first extracted from the nuclei of cells in 1869. Its importance was not recognized for many years. In the early 1950s, two young scientists, James D. Watson (1928–) and Francis H. Crick (1916–2004), studied the molecule. They used imaging data provided by their colleague Rosalind Franklin (1920–1958) to understand the structure of the molecule. They concluded that the DNA molecule is a complex double strand of molecules. It has two "backbones" that run parallel to each other.

Watson and Crick started building a model. Their model began to look like a twisted ladder. The backbones (sides of the ladder) are alternating sugar and phosphate molecules. The steps of the ladder are pairs of four base molecules. The base molecule cytosine (C) always pairs with base guanine (G). The base molecule adenine (A) always pairs with base thymine (T). The sequence of these base pairs encode an organism's genetic information.

Watson and Crick examine their DNA model.

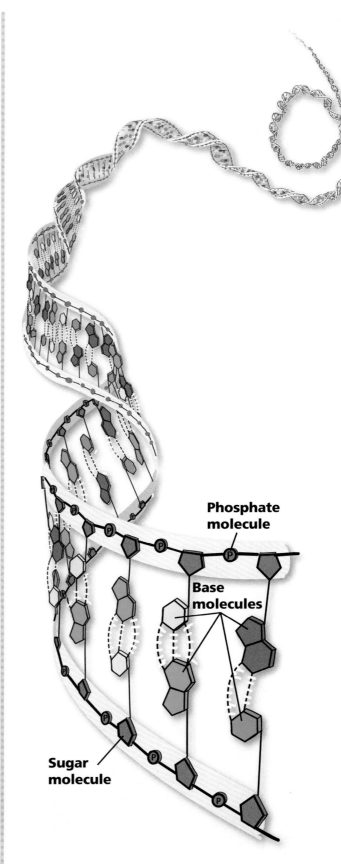

Scientists use the term double helix to describe the twisted ladder shape of a DNA molecule.

DNA

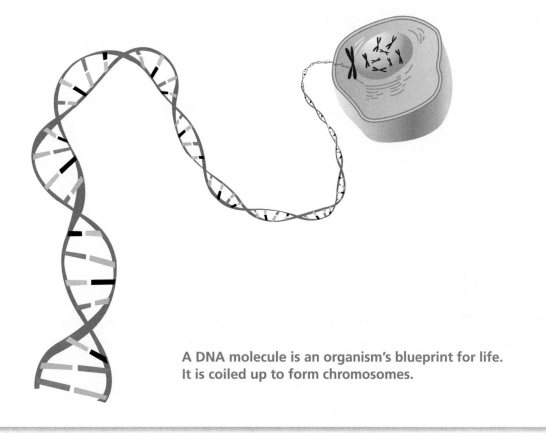

A DNA molecule is an organism's blueprint for life. It is coiled up to form chromosomes.

How does a large "ladder" fit in the nucleus of a cell? The DNA molecule is twisted and twisted into a compact structure. That's what a chromosome is—a twisted DNA molecule.

Genes and Heredity

On a chromosome, sections of DNA, called **genes**, contain coded instructions to make molecules called **proteins**. Humans have thousands of different proteins. Each protein plays a different role. In fact, proteins organize all the other molecules in the human body. They help build cell structures and cells. You could say that genes carry the instructions to make proteins, which determine the traits of an individual.

An organism receives one set of chromosomes from its mother and one set from its father. So it ends up with two copies of every gene, one from each parent. But how does the organism choose which genetic information to use? Well, it doesn't. Mendel's great insight was that an organism's genetic information (its **genotype**), received from its parents, follows an orderly predictable pattern to determine the organism's traits.

The discovery of chromosomes and genes provided a model that explains Mendel's observations. Watson and Crick's model of DNA helped describe how heredity works. Researchers are now working to understand heredity on a more individual level. Can gene sequences be modified? Can inherited conditions be prevented? You will learn more about these developments at the end of the course.

Mendel and Punnett Squares

What was Mendel's breakthrough? What did this Austrian monk observe in his garden that established the foundation of modern genetics and the study of heredity?

Gregor Mendel (1822–1884) was a keen observer of nature. He observed that garden peas have lots of variation from plant to plant. He focused his experiments on several **features** of the pea plant. These included flower color, seed color, and plant height. Mendel chose plants with two possible traits for these features. Some plants produced purple flowers, and some produced white flowers. Some plants produced green seeds, and some produced yellow seeds. Some plants were tall, and others were short. Mendel decided to look for patterns in these traits over multiple generations.

Mendel patiently bred and observed the garden pea plant, *Pisum sativum*. These plants were ideal for his experiments because they grow and reproduce very quickly.

Methods

Mendel designed three stages for his experiment.

Stage 1. Mendel raised several generations of pea plants. He used pollen from a plant to pollinate other flowers on that *same* plant. He planted seeds from the mature plants, raising generations of self-pollinated plants. After a few generations, they bred pure. Breeding pure means all offspring have exactly the same traits as the parent. In Mendel's case, tall plants produced seed that resulted in only tall offspring. Short plants produced seed that resulted in only short offspring. These pure-breeding tall or pure-breeding short plants became the stage 2 parents. Mendel identified them as the **parent generation (P generation)**.

Stage 1: Self-Pollination

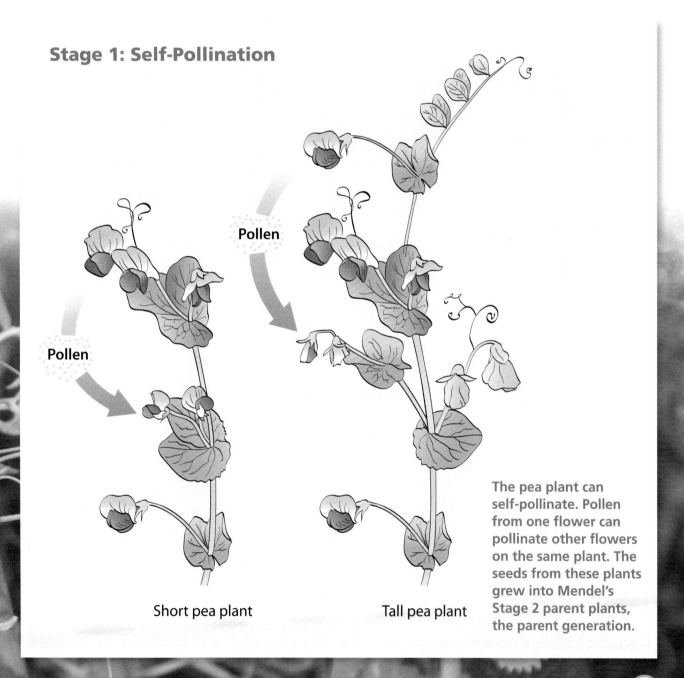

The pea plant can self-pollinate. Pollen from one flower can pollinate other flowers on the same plant. The seeds from these plants grew into Mendel's Stage 2 parent plants, the parent generation.

Investigation 2: Heredity

Stage 2. Mendel carefully cross-pollinated tall parent plants with short parent plants. He placed pollen from tall plants on the flowers of short plants, and pollen from short plants on the flowers of tall plants.

Mendel called the first offspring from the P generation the first **filial** generation. (*Filial* means sons and daughters.) He identified them as the **F_1 generation**. When Mendel cross-pollinated the P generation to produce an F_1 generation, all the F_1 plants were tall. The short trait disappeared.

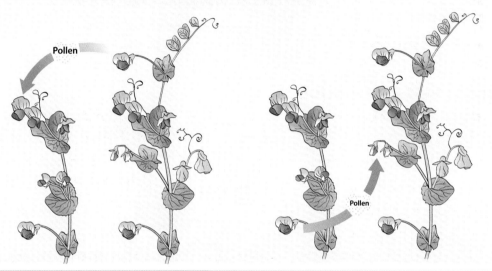

Stage 2: P Generation Cross-Pollination

Mendel prevented natural self-pollination by removing stamens from the flowers of certain plants. He then controlled cross-pollination by dusting pollen by hand between different plants.

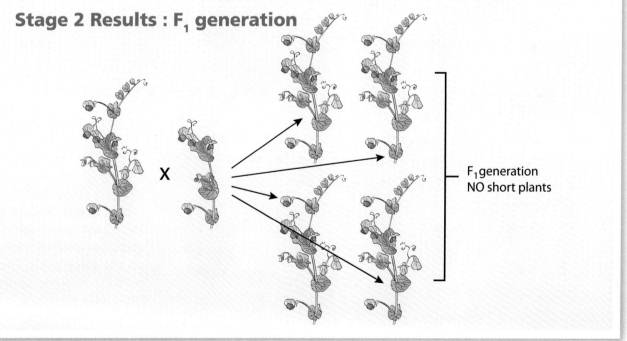

Stage 2 Results : F_1 generation

F_1 generation NO short plants

In Stage 2, Mendel crossed tall and short pure-breeding plants. The results were consistent: only tall plants appeared in the next generation.

Stage 3 Results: F$_2$ Generation

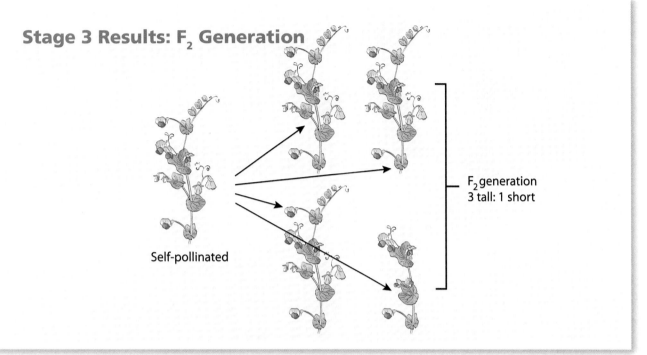

Self-pollinating the plants from the F$_1$ generation resulted in some tall plants and some short plants. The short plants reappeared in the F$_2$ generation.

Stage 3. Mendel pollinated each plant from the F$_1$ generation with its own pollen. The offspring were called the **F$_2$ generation**. When he pollinated all the tall F$_1$ plants with their own pollen, some plants in the F$_2$ generation were tall and some were short! The trait of short height disappeared in the F$_1$ generation and reappeared in the F$_2$ generation.

Here is a diagram of Mendel's results. Count the number of short and tall plants. What ratio do you find in the F$_2$ generation?

When Mendel counted the number of tall and short plants in the F$_2$ generation, he found that the ratio of tall to short plants was 3:1. How could one out of four plants be short, when short plants were absent from the F$_1$ generation?

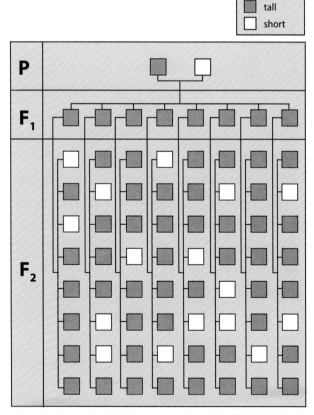

Mendel's painstaking experiments established that traits pass from parents to offspring in mathematically predictable ways.

Investigation 2: Heredity

Children may inherit traits that are expressed in their grandparents but not in their parents. Mendel's observation of dominant and recessive alleles explains how traits can skip a generation.

Mendel's Breakthrough

Mendel had a breakthrough idea. He reasoned that offspring inherit something that determines their traits. He called the inherited thing a factor. These factors are now called genes. Mendel concluded that genes come in pairs. Each member of a gene pair is called an **allele**. An organism has two alleles for each trait, one from the mother and one from the father.

Mendel used the terms **dominant** and **recessive** to describe what he observed. When he crossed a pure tall plant (both alleles for tall height) with a pure short plant (both alleles for short height), all the F_1 offspring received one tall-plant allele and one short-plant allele. Only the tall trait appeared. So he called the tall-plant allele dominant.

Mendel reasoned that the short-plant allele was still there, but was hidden by the dominant tall-plant allele. He called the short-plant allele recessive. The recessive trait would appear only if the offspring inherited the recessive allele from both parents. In the F_2 generation, that combination would happen one time for every three dominant combinations, explaining the 3:1 ratio of tall to short plants.

In 1865, Mendel announced that organisms pass units of information to their offspring during reproduction. This inheritance allows the offspring to develop like their parents. He didn't know what the units were, but he understood how they acted. Without being able to see them, Mendel had discovered the existence of genes and described how they work.

Genes and Punnett Squares

By the early 1900s, chromosomes had been discovered. Walter S. Sutton (1877–1916) was the first scientist to recognize that chromosomes carried Mendel's factors. In 1905, Reginald Punnett (1875–1967), published a textbook called *Mendelism* that introduced **genetics** to the public. He also made a model that predicts the probability of possible genotypes and the **phenotypes** (the appearance of the traits) they produce. Punnett's model is still used. It is a simple two-coordinate system called the **Punnett square**. Let's use Mendel's pea-plant discoveries to see how it works.

Use the letter t to represent the gene that determines plant height. T represents the dominant allele and t the recessive allele.

Mendel's pure tall pea plants had two dominant alleles (TT). The pure short pea plants had two recessive alleles (tt). Both plants were **homozygous** for the height feature. This means the alleles were identical. The tall plant was homozygous dominant (TT) and the short plant homozygous recessive (tt).

Female egg (tall)

T   T

Male pollen (short) t

t

Mendel's published findings from his pea plant experiments were ignored by the scientific community for more than half a century. But the principles he proposed in 1866 are the basis for our understanding of inheritance.

Filling in the squares with the alleles produces four possible offspring. Each offspring has only one possible combination of alleles (Tt). Because all the offspring have a dominant allele, they all have the tall-plant phenotype. The genotypes of these plants are **heterozygous**. The gene is represented by one dominant and one recessive allele.

The result was a ratio of three tall to one short offspring. The Punnett square explains this ratio. Of the four possible offspring, three have the dominant allele (one TT and two Tt), resulting in three tall plants. The fourth has two recessive alleles (tt), resulting in one short plant.

We can **infer** two things. Each offspring of a cross between peas with the Tt genotype has a 75 percent chance of inheriting the tall phenotype and a 25 percent chance of inheriting the short phenotype. In a **population** of peas, the ratio of traits of the plants will be close to 75 percent tall and 25 percent short.

Let's see what happened when Mendel pollinated the F_1 generation of tall plants with their own pollen. Remember that each plant was tall, but had one dominant and one recessive allele (Tt).

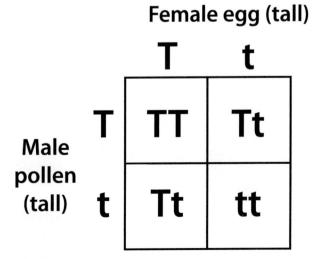

Farmers have long known the usefulness of selective breeding—breeding plants over many generations to achieve certain desirable traits in crops.

Most physical traits, including hair texture and eye color, are inherited traits. Some traits, like dimples, are determined by only one gene. Others, like eye color, are controlled by many genes.

Summary

Mendel's work predicted that traits could disappear in one generation and reappear in the next. His factor idea explained the observations he made in his experiments. Furthermore, he could predict the number of offspring that would have a dominant trait and a recessive trait.

Mendel's experiments uncovered two important principles in the science of heredity. 1) Two factors (alleles) determine traits. One allele comes from each parent. 2) Alleles can be dominant or recessive. Recessive alleles can be present but invisible in an organism.

Punnett's model was based on Mendel's understanding. The Punnett square allows us to calculate the probability that certain genetic combinations will appear in offspring.

Predicting traits in offspring is simple only for traits that are determined by one gene. Mendel was fortunate to choose traits that were determined by only one gene. However, most traits are influenced by many genes.

Investigation 2: Heredity

Mapping the Human Genome

The first eukaryotic organism to have its entire set of genes mapped was baker's yeast, a single-celled fungus. The DNA sequence was released in 1996.

At the close of the 20th century, the scientific community proposed a bold project to map the entire **human genome**. This huge international project would identify the entire sequence of human DNA, billions of base pairs. It was just 150 years earlier that Mendel puzzled over his pea plants. Now the Human Genome Project began.

Every organism has a unique **genome**. A genome is a catalog of the complete sequence of DNA in all of an organism's chromosomes. Remember, it is these sequences that code for the proteins that organize and build the structures of the body. If you could untwist all your chromosomes to look at the base pairs in the DNA molecule, you would be looking at your genome.

An organism has almost exactly the same genome as all other members of its species. For example, more than 99.9 percent of your genome is exactly the same as your mother's, your brother's, your best friend's, and everyone else's on Earth. This common genetic code is called the human genome.

The Human Genome Project

In 1990, the Human Genome Project was launched. This international project was coordinated by the United States Department of Energy and the National Institutes of Health. The goal was to map the entire human genome. Recall that the rungs of the DNA "ladder" are pairs of bases. Cytosine (C) always pairs with guanine (G), and adenine (A) with thymine (T). There are about 3 billion base pairs in human DNA. To sequence the human genome, scientists had to identify each of the 3 billion base pairs. That was a tremendous task.

The project was originally expected to take 15 years. Rapid advances in laboratory techniques and the coordination of many labs sped it up. Sequencing the human genome was successfully completed in April 2003. The work of the Human Genome Project has inspired researchers to learn about the functions of genes and proteins in humans. This knowledge may help us understand the genetic basis for human health, disease, and aging.

Did You Know?

There were two concurrent efforts to decipher the human genome. One was private, the other public. The two efforts joined forces in 2000 to publish their results: essentially all the genes in human DNA.

The human genome is made up of approximately 20,000 genes. The Human Genome Project helps us understand how we humans are all amazingly similar and absolutely unique.

Investigation 2: *Heredity*

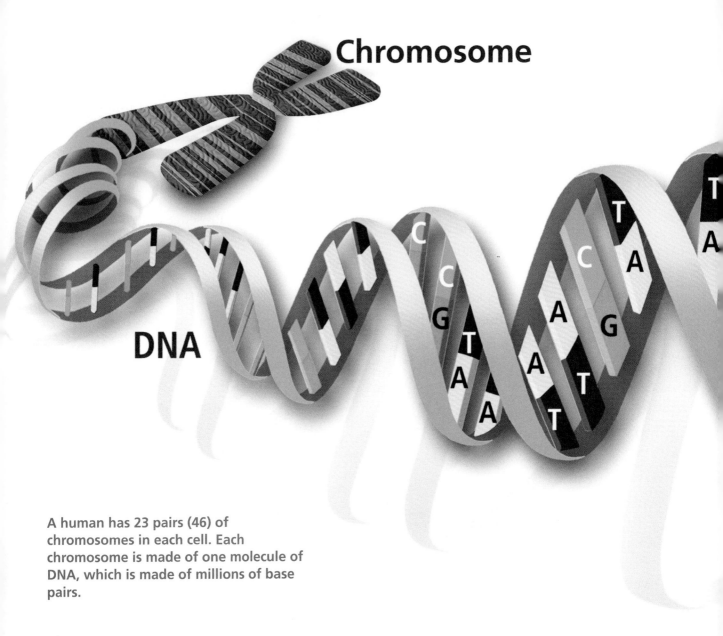

A human has 23 pairs (46) of chromosomes in each cell. Each chromosome is made of one molecule of DNA, which is made of millions of base pairs.

The Human Genome and Disease

Many human diseases are based in our genes. Scientists have discovered more than 1,800 genes associated with single-gene inherited diseases. Studying the human genome can give insight into how to diagnose and treat traits that increase the risk of disease. Comparing human genomes gives clues to why some people are likely to get certain diseases, while others are not.

There are now more than 2,000 genetic tests that help people learn their genetic risk for different diseases. Some of these tests involve asthma, heart disease, diabetes, dementia, and cancer. Having a high risk, however, does not mean that you will get that disease. Let's look at an example.

Autosomal Dominant Inheritance

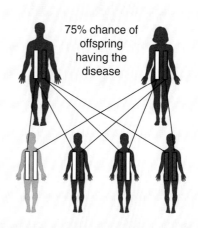

Autosomal dominance is a single-gene inheritance pattern. This kind of disease or disorder can be passed to offspring even if only one parent has the abnormal, or mutated, gene. Huntington's disease is caused by a single-dominant defect gene.

A gene called BRCA controls normal cell growth in breast tissue. A **mutation** is an error in the DNA sequence. Two BRCA mutations (BRCA1 and BRCA2) have been linked to increased chances of developing breast cancer. The BRCA gene test can determine if a person has one of these mutations. If that person does, he or she has a higher likelihood of developing breast cancer.

What does this information do for at-risk women, and more rarely, men? This knowledge can help at-risk people by detecting cancer sooner with frequent testing. Earlier treatment often leads to a longer and healthier life. But this information could also make things more difficult. Even if someone does have one of the mutations, there is a good chance that he or she will not develop breast cancer. For a healthy person, the fear of developing cancer could lead to unnecessary medical procedures.

Scientists are using the information in the human genome to develop drugs that target specific genes. In the future, such drugs might be powerful tools to help prevent or treat genetic diseases.

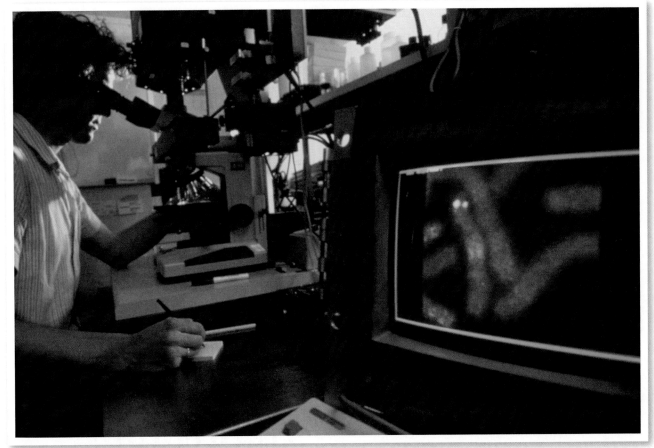

Advances in technology make it possible to view cells and organisms up close.

Ethical Considerations

Researchers have mapped the genomes for many organisms, including microorganisms, plants, and animals other than humans. They use this information to model biological systems to solve health and environmental challenges.

As with many scientific advances, however, genome information brings up ethical questions. Here are some questions that are being discussed.

- Who should have access to genetic information about a person?
- How should families, employers, and medical professionals act on genetic information?
- Should we allow genetic manipulation and therapy, or should it be prohibited?
- How might having genetic information about yourself change your life?

Take Note

You will look more closely at genetic technologies at the end of this course. What are your current ideas about using genetic information in medicine?

Adaptation

Odds are that birds native to where you live are very different from the flamingo. The flamingo is adapted to salty wetland environments.

Birds are found almost everywhere in the world. Do the birds where you live look anything like this? Take a look outside. With any luck, you'll see several kinds of birds.

Flamingos have the longest legs and necks, relative to body size, of any bird. They can also drink hot salty water. How do these adaptations help them live in their environment?

Adaptations for an Extreme Environment

Let's travel to a totally different environment than the one you live in—Antarctica. Antarctica is the windiest, coldest, and driest continent on Earth. The South Pole is located in Antarctica. About 98 percent of the continent is covered with ice. Antarctica harbors 45 species of birds during the year. What **adaptations** allow birds to survive in this extreme environment? An adaptation is an inherited trait or behavior that increases an organism's chances of surviving long enough to reproduce.

Take Note

What physical characteristics and behaviors help birds survive and reproduce in Antarctica? List your ideas in your notebook.

Penguins lost the ability to fly as their wings changed over time into flippers that help them "fly" through the ocean.

Consider two birds that live in the Antarctic, the Adélie penguin and the wandering albatross. Take a close look at their beaks. How might their beaks help them obtain food and raise their young? Look at their wings and body color. How might these traits help them survive? What behaviors do you think could help them survive and reproduce? Let's find out more about each of these birds.

Both Adélie penguins and wandering albatrosses range throughout the polar region of the southern ocean. Beyond that, they have little in common.

The wandering albatross has the longest wingspan of any bird, allowing it to ride the winds and glide for hours, even days.

Investigation 3: Evolution

An Adélie penguin can toboggan faster than a running human for short distances! The penguin propels forward with its feet and uses its flippers for balance and as oars.

Adélie Penguins

Adélie penguins don't fly, but they are excellent swimmers. They can swim 150 kilometers (km) to reach food for their chicks.

Penguins have very short legs, but they can walk long distances. In the early spring, they might walk 50 km across the pack ice. When they find open sea, they gather food for their chicks. They can also "toboggan." Tobogganing is sliding on their bellies and pushing with their feet. This saves energy on long trips across the ice.

Penguins are not very graceful walkers.

Adélie penguin nests are depressions in the snow lined with small stones, which keep the eggs free from melt-water.

Adélie penguins breed on rocky areas along the Antarctica coast. Both parents take turns incubating the eggs during the "warm" months of October through February. They also take turns guarding the chicks and getting them food. When mature feathers replace their down (this happens around 9 weeks of age), the chicks are ready to go to sea to feed themselves. They eat fish and krill, small shrimp-like crustaceans that live in the ocean.

Warming Antarctic seas are causing populations of krill to decrease, affecting the animals that use it for a food source.

Investigation 3: *Evolution* **45**

Penguins are at greatest risk from predators when entering or leaving the water to hunt or migrate. They jostle for position until one finally leaps or is pushed off the edge.

Adélie penguins spend the winter at sea in the pack ice. They can swim 1,200 km away from their breeding sites. They gather in large groups before jumping into the sea. This increases an individual's chance of escaping predators. Their main predators are the leopard seal and the skua. The skua is a bird that steals penguin eggs and young chicks. The 1 metric-ton leopard seal can easily kill any penguin that doesn't stay with the group.

What other adaptations help Adélie penguins survive? Their compact shape reduces heat loss. Dense feathers and a layer of fat insulate them. Their black-and-white feathers make it harder for predators to see them when they are in the water and help with warming and cooling on land. Their heart rate slows during deep dives, which helps them efficiently manage oxygen use. Their salt glands and kidneys allow them to drink sea water.

One disadvantage of huge wings is difficulty taking off. Albatrosses usually nest on windswept heights, making a launch much easier.

A young albatross can be identified by its dark feathers. It fledges, or develops the wing feathers necessary for flight, at about 7–9 months.

Wandering Albatross

Just like the Adélie penguin, the wandering albatross has adaptations that allow it to survive in the extreme environment of Antarctica. But as you read, note how different the two birds are.

Albatrosses can fly long distances because of the structure and length of their wings. Their wingspan is about 3.5 meters (m)! Their shoulders lock the wings in an outstretched position. This lets them fly on wind currents without using much energy. They sometimes soar for several days, traveling thousands of kilometers to find food.

The wandering albatross has two long tubes along its bill. These tubes give it a great sense of smell. The albatross can smell fish and squid from many kilometers away. It hunts at night, making shallow dives to catch its prey.

An albatross soars at speeds averaging 40 km per hour, often following ships and fishing boats for discarded food. In fact, the chief hazard to the albatross is longline fishing.

Investigation 3: Evolution **47**

Wandering albatrosses nest on several islands in the Antarctic Ocean. Their nesting sites are open ridges with some vegetation. They breed every other year, and produce only one egg at a time. The male and female take turns incubating the egg. They spend more than a year raising their chick.

What other adaptations help wandering albatrosses survive? They have large webbed feet for paddling. Lots of melanin in their dark flight feathers slows deterioration from salt and sunlight. They can produce an energy-rich stomach oil from the organisms they eat. This oil is an efficient source of food for chicks and gives the adults energy for their long flights.

Albatrosses are rarely seen on land, gathering in colonies on remote islands only to breed.

Wandering albatrosses mate for life. Though long-lived, these birds start breeding late and reproduce at a slow rate. As a result, many populations are in long-term decline.

How Do Adaptations Arise?

The wandering albatross and the Adélie penguin are marvelously adapted to their environment. How did they acquire the adaptations that allow them to survive and reproduce? To answer that, we need to consider mutations and DNA.

Mutations are the source of adaptations. A mutation is any error in an organism's DNA. If an error occurs in the DNA of an egg or sperm cell, that error will pass on to the offspring. Remember that DNA codes for proteins, which direct the structure and function of an organism. If a mutation causes a change in proteins, the organism's traits may change. If the new trait is beneficial, it may help the organism survive longer and have more offspring. Those offspring will also have a greater chance of survival and will continue to pass on the revised DNA sequence to their offspring. The genetic mutation has resulted in an adaptation.

The blue-and-yellow macaw is native to tropical areas in South America. Based on its sharp, hooked beak, what kind of food do you think it eats?

All organisms have adaptations that allow them to survive and reproduce. Adélie penguins live in a cold, treeless, icy environment. Their food comes from the sea. Their behavior and body shape allow them to be successful and thrive in that environment. Compare the beak of the penguin to that of the blue-and-yellow macaw shown above.

The penguin eats fish, and the macaw eats nuts and fruits. Their beaks are adaptations that make them successful in their environments. Think about a crow or a pigeon or any other bird where you live. Their behavior and body shape are not adapted to the Antarctic environment, and a penguin is not adapted to where you live. Penguins can exist only in zoos in the United States. Nor would you find a crow or pigeon in Antarctica—it could not survive.

All those adaptations originated in genetic mutations in the ancestors of these birds. Over time, they led to new species of birds.

Changing Environments

The fossil record provides clues about why penguins became better swimmers than flyers. Their abilities shifted when other species of animals grew more dependent on food in the sea and competition for food increased. Mutations that allowed for better swimming and diving would have enabled some penguins to survive and reproduce, eventually leading to adaptations in their wings and body shape. At the same time, these changes compromised their ability to fly.

Right now, climate warming is decreasing the sea ice in Antarctica. Organisms such as krill, a major food source for penguins, feed on algae growing on the underside of the ice. When sea ice shrinks, the growth of ice algae shrinks. This affects the populations of krill, which affects the population of penguins. Adélie penguins are adapted to sea ice. As the sea ice shrinks, populations of Adélie penguins will shrink. They will be replaced by penguin species that are better adapted to open water.

Changes to the sea ice, or frozen ocean, surrounding Antarctica may threaten species that live in that extreme environment.

Investigation 3: Evolution

Unlike sea ice, shelf ice is fresh water. Ice shelves are the floating masses of ice that hug the Antarctic coastline. Warming temperatures are causing ice shelves to collapse and disintegrate at a rapid pace.

Adaptations develop over a long time. If the climate changes quickly, mutations and adaptations will not keep pace. What kinds of variation in Adélie penguin structure or behavior might help them survive a changing environment?

Adélie penguin parents share egg and chick duties. How is this behavior an advantage for survival?

Think Questions

1. List one physical trait and one behavioral trait of the Adélie penguin that help it survive in the Antarctic. Would those two traits be an advantage for survival in the environment where you live? Explain.

2. List one physical trait and one behavioral trait of the wandering albatross that help it survive in the Antarctic. Would those two traits be an advantage for survival in the environment where you live? Explain.

Natural Selection

In November 2011, Ken Koehler was violently ill. He was kneeling on the floor of his bathroom, vomiting and suffering from intense stomach pain.

A friend rushed him to the emergency room. Doctors tracked down the culprit, the bacterium *Salmonella typhimurium*. They gave Koehler antibiotics to kill the bacteria. The salmonella was resistant to the first several drugs they tried. The doctors finally found a different kind of powerful antibiotic that worked, and Koehler recovered.

The Centers for Disease Control and Prevention tested the bacteria that caused Koehler's illness. They traced the bacteria to ground beef. The bacteria were resistant to nine common antibiotics. These antibiotics normally kill *S. typhimurium*, but they no longer worked. How did the bacteria become resistant to standard antibiotics? The answer lies in adaptation and the process of **natural selection**.

> **Take Note**
>
> **Look for the definition of adaptation in your science notebook.**

Eating undercooked burgers is dangerous! The only way to be sure meat is cooked to a temperature high enough to kill harmful bacteria is to use a meat thermometer.

Investigation 3: Evolution

Process of Natural Selection

Adaptations result from mutations in an organism's DNA. Sometimes a mutation produces a trait that helps the individual survive. That individual may have more offspring than other individuals in the same population. Its offspring are likely to inherit the trait. With each generation, the trait will become more common in the population. This is the essence of natural selection. Natural selection occurs whenever

- variation of traits exists in a population,
- the traits are inherited, and
- individuals with beneficial traits for a particular environment reproduce more than other individuals in a population.

What variation of traits do you observe in this pack of wolves? How might you explain the variation?

Larkeys

You have seen natural selection in the larkey population that moved from the mountains to the prairies.

Natural selection was the mechanism that caused the change in the prairie larkey population. Long-legged larkeys survived to produce more young. So long legs became more common than short legs. The successful long-legged larkeys passed the gene for long legs to their offspring. After many generations, the short-legged trait could even disappear.

What larkey traits are best adapted to the prairie?

Take Note

What information did you gather from the larkey simulations that provides evidence to support natural selection?

A genetic mutation in female blue moon butterflies is passed on to males who then survive bacterial infection.

Blue Moon Butterflies

Let's look at another example of natural selection at work. The beautiful blue moon butterfly lives on a few tropical islands in the South Pacific. The butterflies are often infected by *Wolbachia* bacteria. These bacteria kill the female butterfly's male embryos. Very few males are produced. You can imagine what might happen to an infected population of butterflies. It probably would not survive for long. In 2007, however, scientists reported a dramatic change. The percentage of males in one population jumped from 1 percent to about 39 percent within ten generations. What happened?

The scientists discovered a genetic mutation in female butterflies. The males that inherited this mutation were resistant to the *Wolbachia* bacteria! Resistant males survived to mate with many females. This quickly spread the resistance trait throughout the population.

Scientists regard the comeback of the male blue moon butterfly as possibly the fastest evolutionary change ever monitored.

These are some questions the scientists discussed. Discuss them with a partner.
1. Is there variation in the blue moon butterfly's ability to resist the *Wolbachia* bacteria? Where did the variation come from?
2. Is resistance an inherited characteristic? What is your evidence?
3. Is resistance a positive adaptation?
4. What happened to the population of blue moon butterflies over several generations?

Researcher Gregory Hurst said, "We usually think of natural selection as acting slowly, over hundreds of thousands of years. But the example in this study happened in a blink of the eye, in terms of evolutionary time."

Take Note

Could *Wolbachia* evolve and become able to counter the resistance built up by the butterflies? What do you think?

Investigation 3: Evolution

Salmonella typhimurium

Now let's return to the antibiotic-resistant *Salmonella typhimurium*. Salmonella bacteria are found in cattle. If a cow is sick, it gets antibiotics to treat its infection. However, several antibiotics are often given to cattle to *prevent* illness and to make them grow faster. Many of these antibiotics are the same ones given to people. Random mutations in the bacterial genome enable some bacteria to survive the antibiotics. Bacteria reproduce rapidly, sometimes doubling their number in as little as 20 minutes. Because bacteria with antibiotic resistance are not killed by the antibiotics, they quickly produce a large population of antibiotic-resistant bacteria.

When Ken Koehler handled the ground beef, antibiotic-resistant bacteria transferred to his hands. He touched other food that he ate. The bacteria traveled to his intestines and made him sick. Doctors prescribed the usual antibiotics for a salmonella infection. The antibiotics failed to kill the bacteria because they had become resistant, due to natural selection. The antibiotic-resistance trait helped them survive.

Mutations in Bacteria

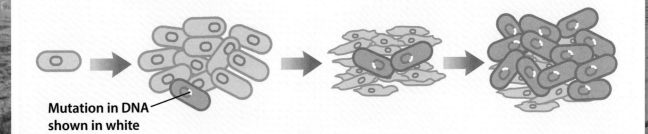

Mutation in DNA shown in white

Mutations in bacterial DNA can lead to antibiotic resistance.

Summary

Because environments are constantly changing, organisms must also change if they are to survive and thrive. Natural selection is the mechanism that drives these changes. We have seen this in the camouflage of rock pocket mice, antibiotic resistance in bacteria, disease resistance in butterflies, and even in larkey leg length.

Think Question

Is it possible for a new species to evolve from an existing one as a result of natural selection? Explain your thinking.

Salmonella-tainted beef may be one of the toughest challenges facing the cattle industry. The drug-resistant nature of the bacteria gives researchers few prevention options.

What Makes a Scientific Theory?

In a conversation with a friend, you might say, "You know, I have a theory that Jon has a crush on Ava." But our everyday use of *theory* is different than its scientific meaning.

Your **theory** about Jon and Ava might be a hunch, with little supporting evidence or broad application. In science, however, a theory is a well-tested explanation. It makes sense of many facts, observations, and hypotheses. It allows scientists to accurately predict what they will observe. Scientific theories are testable. Often scientists find new evidence that discredits part of a theory. They revise the theory to include the new evidence.

A whispered rumor may be referred to as a theory, but it is more likely just gossip. In contrast, a scientific theory is backed up by tested evidence.

The Theory of Plate Tectonics

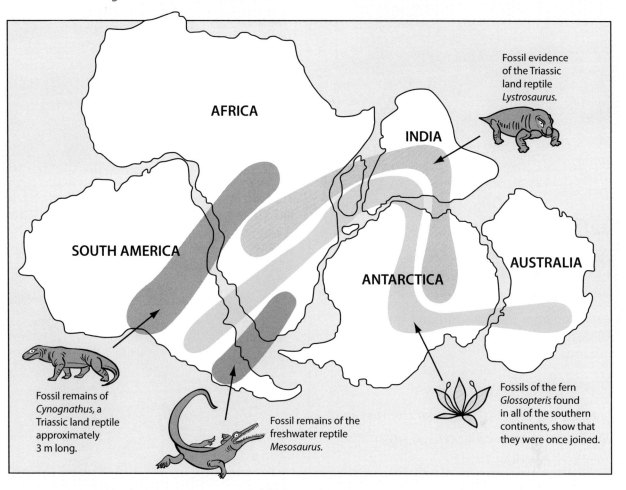

The distribution of fossil evidence supported the idea that continents were once connected and then moved apart. This contributed to the theory of plate tectonics which explains patterns of earthquakes and volcanoes.

How did your theory about Jon and Ava hold up? Your friend tells you that Jon and Ava are brother and sister. They often walk home together. The new evidence provided by your friend reveals that your theory was an incorrect conclusion based on misinterpreted observations.

Scientific theories stand up over time. Here is a list of well-tested, long-lived theories.

- The theory of plate tectonics explains the movement of huge plates beneath Earth's surface.
- Atomic theory explains the particle nature of matter.
- The theory of gravity explains why we don't fly off Earth.
- Germ theory explains how bacteria and other microorganisms cause disease.
- The **theory of evolution** explains why life is both similar and diverse.

Investigation 3: Evolution **61**

Darwin compared the South American rhea with distinct but similar species on other continents.

The ostrich is native to Africa. It cannot fly, but like the emu and rhea, it is a very fast runner. It can run for kilometers at speeds of over 15 km per hour.

Observing Patterns

Let's look more closely at how the theory of evolution has become the central organizing principle of biology.

Charles Darwin (1809–1882) was 22 years old when he sailed on an expedition to map coastlines around the world. For 5 years, Darwin collected specimens and made drawings and notes about the organisms and ecosystems he observed. He noted three patterns: species vary globally, species vary locally, and species change over time.

Species vary globally. Darwin noticed that similar ecosystems around the world have species with similar traits. For example, he noticed similarities among the rheas of South America, the ostriches of Africa, and the emus of Australia. Yet they are clearly different species. Darwin wondered how similar traits in different species could be so widespread.

The emu is the largest native bird in Australia.

Species vary locally. Darwin noticed many related species living in the same environment. For example, several species of tortoise and several species of mockingbirds coexisted on the Galápagos Islands. The most famous example was the many species of finches there. Darwin wondered how so many similar species could survive in the same place.

Species change over time. Darwin collected many fossils. As his collection grew, he noticed that the fossils of some extinct animals had traits like those of animals living today. For example, the long-extinct Glyptodon looks like the modern armadillo. Darwin wondered how similar-looking species could exist so many thousands of years apart.

Cocos Island finch
Pinaroloxias inomata
(insect eater)

The slender pointed beak of the Cocos Island finch is specialized to eat insects.

Glyptodon, whose name means "carved tooth," was a huge, carapace-covered mammal. It was probably hunted into extinction by early humans about 10,000 years ago.

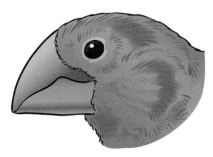

Large ground finch
Geospiza magnirostris
(seed eater)

The beak of the large ground finch can crack open large, hard seeds that other birds cannot.

Compare the modern armadillo to the extinct Glyptodon. How are they similar? How are they different?

Seeking Confirmation and Explanation

After his voyage, Darwin searched for signs of how traits pass from one generation to the next and how traits change in the process. He observed that people have bred many varieties of plants from one species. For example, brussels sprouts, cauliflower, broccoli, and kale were all bred from the same wild mustard plant. He talked to pigeon breeders about their work. Their birds became so different from one another that they appeared to be different species.

Brussels sprouts are *Brassica oleracea* bred for enlarged leaf buds.

Many of the green vegetables that are so good for us are domesticated versions of the same species, *Brassica oleracea*, the wild mustard plant.

Darwin was struck by the variation among domestic rock pigeons, a single breed. We now know of more than 300 pigeon species, including this band-tailed variety native to North American woodlands.

The beautiful fantail pigeon has been bred to have more feathers in its tail than other breeds.

All breeds of pigeon are descended from the rock pigeon.

Darwin spent years reflecting on his observations. He gathered information from other biologists and geologists. In the end, he wrote about a natural process that leads to the survival of organisms best adapted to cope with the demands of their environments. Well-adapted organisms have a greater chance of leaving more offspring than less well-adapted organisms. Darwin called this process natural selection. He realized that natural selection occurs under these circumstances.

- More individuals are born than can survive.
- Traits pass from generation to generation and vary within a population.
- Different traits lead to different rates of survival and reproduction.

Publication

In 1859, Darwin published *On the Origin of Species by Means of Natural Selection*. He proposed that over many generations, successful adaptations and reproductive isolation would lead to new species. He also proposed that today's species are descended from common ancestors. They have changed over vast periods of time. This was the first publication of the theory of evolution. Natural selection is the chief mechanism that leads to evolution.

Many scientists criticized Darwin's ideas. First, at that time, Earth was thought to be no more than several thousand years old. That brief time is too short for natural selection to do its work. Second, fossil evidence was too limited to show the connections between species. Both of these criticisms have been answered by evidence discovered since Darwin's time.

Darwin admitted that he could not explain how traits pass from one generation to the next. Later, Gregor Mendel's evidence of heredity was merged with Darwin's mechanism of natural selection. At that point, the theory of evolution became widely accepted throughout the scientific community.

Revising the Theory

Since Darwin's publication of *On the Origin of Species*, discoveries in many fields have supported natural selection as one of the driving forces of evolution. The theory has been revised to include the new evidence. But Darwin's bold ideas and careful gathering of evidence helped establish the theory of evolution. And that theory organizes our understanding of life on Earth.

Eugenie Scott (1945–), former executive director of the National Center for Science Education, summed up the theory's value. "Theories are actually very important. If we say 'the theory of evolution,' we are praising it. We're not saying it's a guess or a hunch. We're saying it's a very important explanation that helps us understand the natural world better. Theories explain laws and facts. They're some of the most important accomplishments of science."

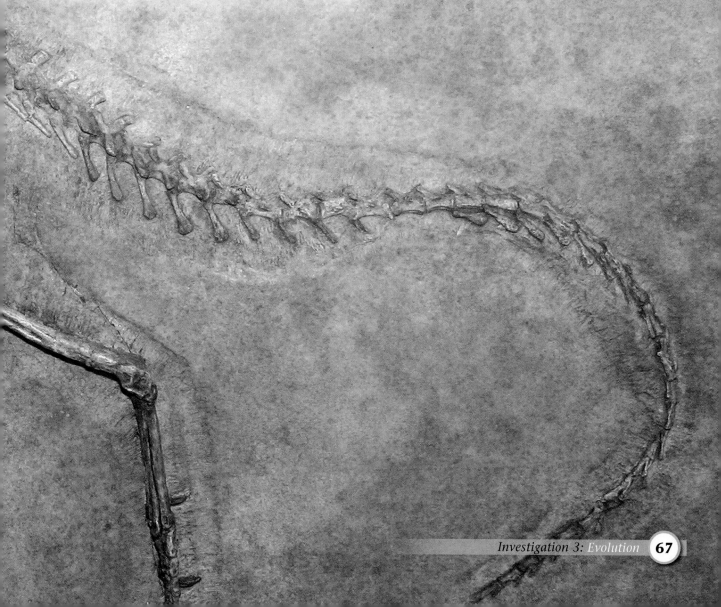

The fossil record now provides much evidence of how species are related and have changed, or even disappeared, over time.

Describing a Scientific Theory

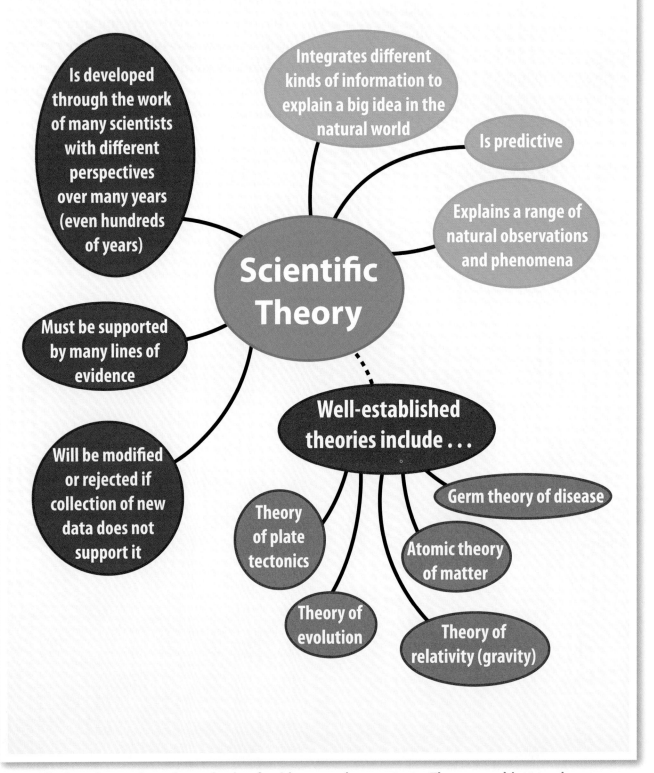

Scientific theories are based on a body of evidence and many tests. They are subject to change as new evidence is discovered or as existing evidence is reinterpreted.

Images and Data

Images and Data   Table of Contents

Investigation 1: The History of Life
Mass Extinctions 73
Transitions 78

Investigation 2: Heredity
A Larkey Yammer 82

Investigation 3: Evolution
Influencing Evolution 84

References
Science Safety Rules 89
Glossary 91
Index 94

Mass Extinctions

The fossil record shows millions of species appearing and disappearing throughout Earth's history. When the last organism of a species no longer exists, we say the species has gone extinct. Scientists estimate that about 99.9 percent of all species that have ever existed on Earth are now extinct.

Species go extinct at a low rate all the time. But Earth has also seen a few mass extinctions. During a mass extinction, more than 75 percent of all species go extinct within several million years. Several million years is a relatively short part of Earth's 4.6 billion years. Evidence from the fossil record points to at least five mass extinctions. What does it take to wipe out most of the life on the planet?

Organisms whose structures include hard parts, like bones and shells, are most likely to leave their mark in the fossil record. These began to appear about 540 million years ago.

The Big Five Mass Extinctions

The Ordovician Extinction

(440 million years ago)

By the end of the Ordovician extinction, 86 percent of all species had died out. That means for every 15 species alive before the event, 13 species went extinct!

Cause: A rapid drop in global temperatures and climate patterns was the likely cause of the Ordovician extinction. Many species could not survive the cooler conditions. Because more water was stored in growing glaciers and ice fields, sea levels dropped quickly. Many marine organisms lost crucial coastal habitat.

Organisms hardest hit: Corals and deep-shelf bottom-dwelling species such as trilobites

| Cambrian | Ordovician | Silurian |

Millions of Years Ago

500   450

The Devonian Extinction

(360 million years ago)

The Devonian extinction lasted 20 million years. During this time, 75 percent of all species died out.

Cause: The fossil record shows a dramatic increase of plant diversity on land just before the Devonian extinction. The many plants took in a lot more carbon dioxide from the atmosphere. Big decreases in carbon dioxide dramatically cooled the world, including the ocean.

Organisms hardest hit: Corals, brachiopods, armored fish, and small-shelled organisms called foraminifera

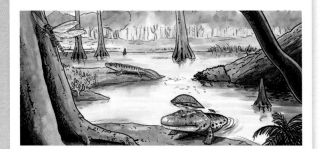

The Permian Extinction

(250 million years ago)

The Permian extinction was the largest mass extinction in Earth's history. In less than 2.5 million years, 96 percent of all species disappeared. All of today's life on Earth is descended from the surviving 4 percent!

Cause: Thick layers of ash in the Permian fossil record imply that there were major volcanic eruptions. Gases from the eruptions probably caused rapid global warming and dramatic changes to ocean chemistry. The life that evolved from the few survivors was very different from earlier life.

Organisms hardest hit: Corals, early terrestrial tetrapods, insects, and trilobites

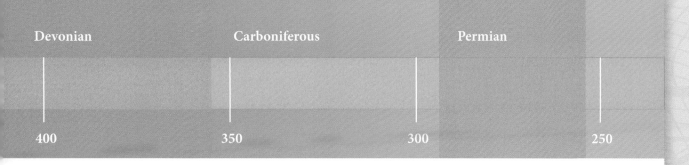

Devonian | Carboniferous | Permian

400 | 350 | 300 | 250

Investigation 1: The History of Life

The Triassic Extinction

(200 million years ago)

The Triassic extinction lasted 1 to 8 million years. During this time, 80 percent of all species went extinct.

Cause: Shifting continents probably led to massive volcanic activity in the Triassic period. Gases released by the eruptions led to rapid global warming and dramatic changes to ocean chemistry. Changing sea levels disrupted coastal marine habitats like coral reefs. Dramatic climate changes killed many land-plant and animal species.

Organisms hardest hit: Amphibians, reptiles, corals, and ammonites

Triassic   Jurassic

250   200

The Cretaceous Extinction

(65 million years ago)

The Cretaceous extinction happened at the end of the Cretaceous period and the beginning of the Tertiary period. The extinction lasted up to 2.5 million years. During this event, 76 percent of all species became extinct, including the dinosaurs.

Cause: Fossil evidence shows widespread volcanic activity affecting life at the end of the Cretaceous period. One opinion is that an asteroid impact in southern Mexico led to a catastrophic event. The fossil record shows a band of ash over the entire Earth. Poisonous gases and dust hid the Sun. Without sunlight, plants died and the temperatures dropped. Without food, large dinosaurs died. Only small creatures survived in the new climate.

Organisms hardest hit: Plants, dinosaurs, large mammals, and mollusks

The Sixth Extinction?

For the first time in Earth's history, one species may be causing a mass extinction. About 10,000 years ago, human populations increased while the rate of extinction of other species dramatically increased.

In the past, changes to large Earth systems caused global climate changes. The current changes are a direct result of human activity. Activities such as burning fossil fuels, pollution, habitat destruction, and overfishing are profoundly impacting life on Earth. Scientists estimate that more than half the world's species could be extinct by 2100.

As the human population grows, we must figure out how to minimize our effect on climate and habitat for other species. Efforts to reduce carbon emissions are very important.

Humans are aware of what is happening to Earth. Earth's future is in our hands!

| Cretaceous | Paleogene | Neogene | Quaternary | Anthropocene |

Future

100   50   0

Investigation 1: The History of Life

Transitions

Ancient Lungfish

Fossils are 390 million years old

Defining characteristics:
- Completely aquatic
- About 35 cm long

Fossil

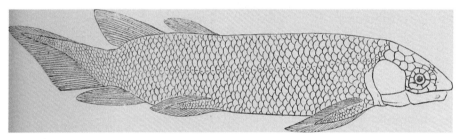
Drawing

Model

Bones in the Ancient Lungfish Fin

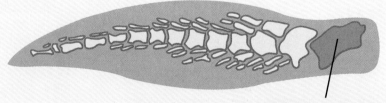

The fin has one bone at the base of the fin.

Eusthenopteron

Fossils are 380 million years old

Defining characteristics:
- Completely aquatic
- Fins have a mix of amphibian and fish characteristics
- Considered a "lobe-finned" fish
- About 1.5 m long

Fossil

Model

Bones in the Eusthenopteron Fin

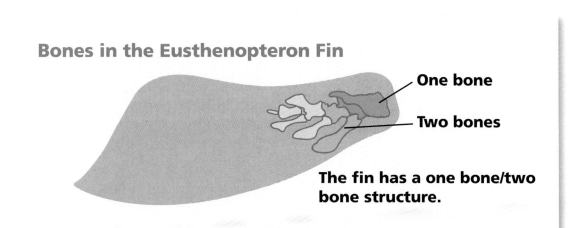

One bone

Two bones

The fin has a one bone/two bone structure.

Investigation 1: The History of Life

Acanthostega

Fossils are 360 million years old

Defining characteristics:
- Paddle-like limbs with digits (first true fingers and toes)
- No defined wrists or ankles
- Primarily aquatic
- About 60 cm long

Note the digits

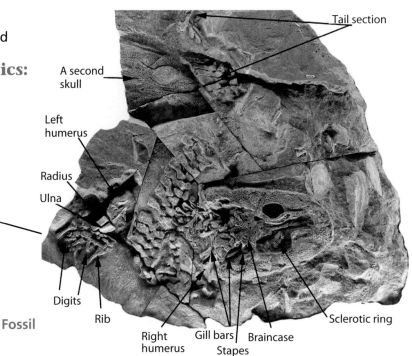

Fossil

Model

Bones of the Acanthostega Paddle

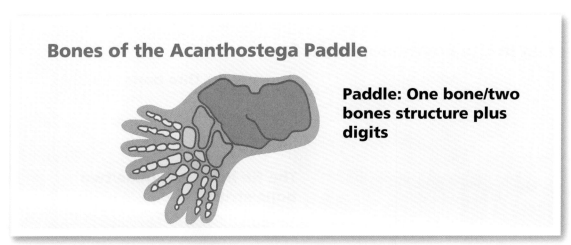

Paddle: One bone/two bones structure plus digits

Pederpes

Fossils are 350 million years old

Defining characteristics:
- True feet with fingers and toes
- Considered the first true terrestrial tetrapod
- About 1 m long

Fossil

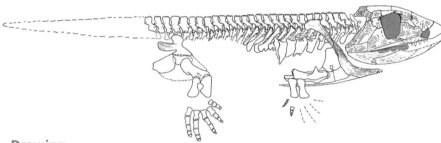
Drawing

Model

Bones of the Pederpes Limb

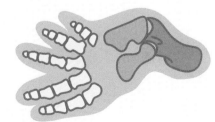

Limb: One bone/two bone/many bones structure plus digits ("many bones" missing from fossil)

Investigation 1: *The History of Life* **81**

A Larkey Yammer

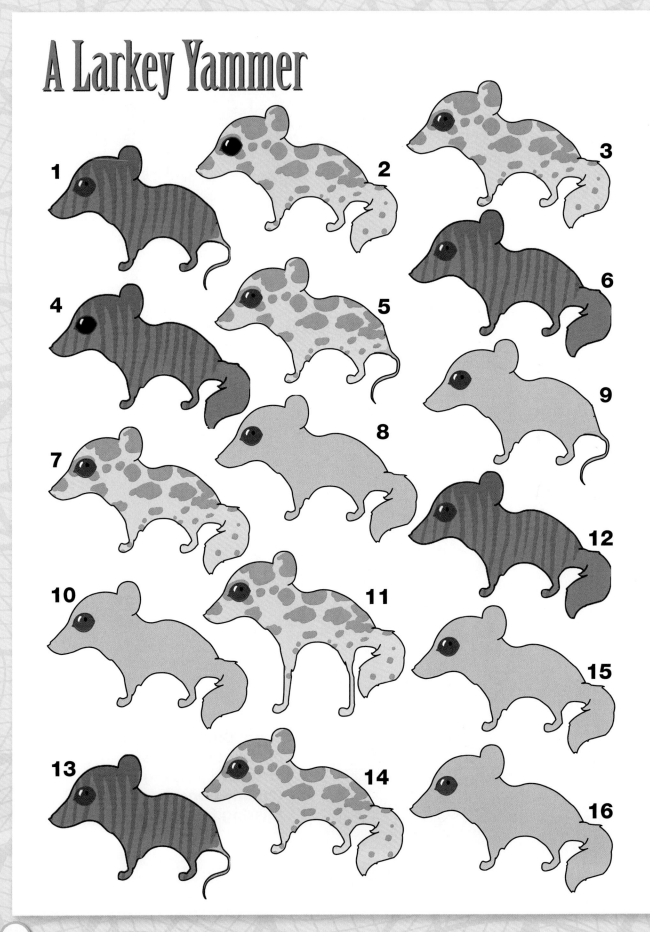

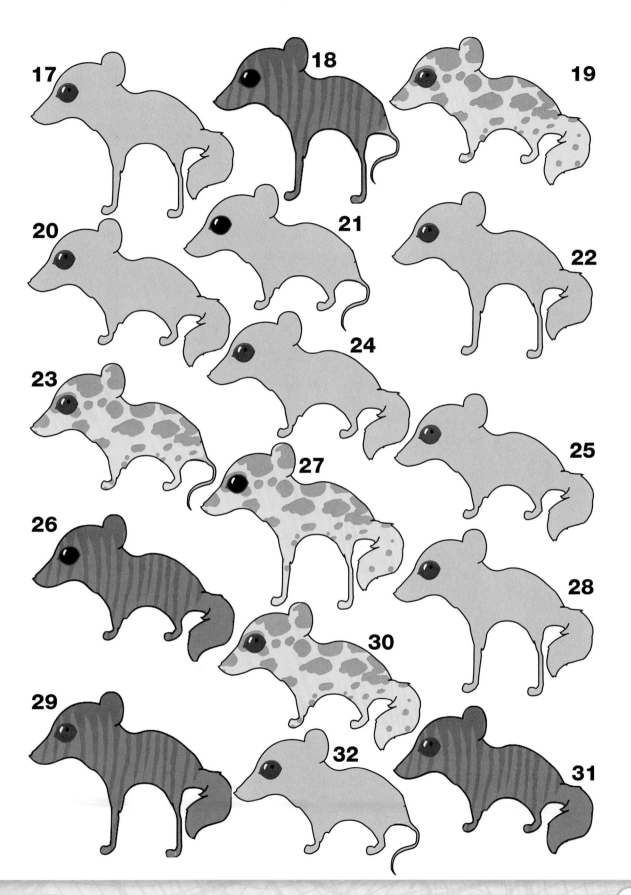

Investigation 2: Heredity

Influencing Evolution

Humans are the only organisms that have the ability to use technology to influence inheritance. For more than 10,000 years, humans have affected the inheritance of plants and animals. They have made useful traits more common.

Humans have changed the genetic makeup of other organisms in several ways. These brief descriptions will help you choose one method or technology to study further.

Genetics researchers explore a wide range of issues with important health implications and ethical challenges. Current topics include stem cell research, DNA profiling, inheritance patterns, and the genetics of infectious diseases.

Genetic Technology

Artificial selection, or selective breeding, requires no modern technology. People decide which traits are desirable, find individual organisms that have those traits, and breed them. For example, humans have selectively bred honeybees for thousands of years. Ancient Egyptians were the first known beekeepers. The common honeybee in the United States today is the European honeybee. Beekeepers in Europe kept their bees in containers. There, they could control which bees mated. The traits they selected led to a honeybee that is docile and resistant to disease. It can produce large quantities of honey.

Humans have long used selective breeding of honeybees to improve the honey harvest. Another important reason to manage honeybees is because humans rely on their pollination activity.

About one-third of our diet comes from insect-pollinated plants, with honeybees the number one crop pollinator by far.

Investigation 3: Evolution

Genetic engineering is a technology that transfers genes between organisms. You can think of it as cutting genes from one organism and pasting them into the genome of another. Scientists first used this technology in the early 1970s when they introduced a piece of DNA from one virus into another. Two ways genetic engineering is used are to develop **genetically modified organisms (GMOs)** for food and to create **transgenic organisms** that produce a needed substance.

GMOs are made when genetic material is introduced into an organism that humans use for food. Soybeans are an example of a food that is commonly altered. Soybean seeds are genetically modified with genes from a bacterium so that weed-killing chemicals do not kill them. Farmers can spray chemicals to kill weeds without injuring the soybean plants.

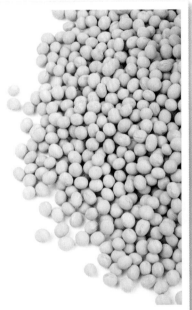

Soybeans are one of the most important crops worldwide for producing protein-rich foods and oils.

More than 90 percent of cultivated soybeans in the United States are genetically modified to resist weed-killing chemicals.

Transgenic organisms have altered DNA. Scientists add a gene from another species to create a transgenic animal. Transgenic technology can create single-celled organisms or modify plant genomes to produce simple proteins such as insulin. Transgenic livestock can produce complex human proteins. These proteins can be used to make medicine, such as blood-clotting factors for people with hemophilia. The chosen piece of DNA is injected into a fertilized egg. The fertilized egg divides as the animal develops, passing the modified DNA to each new cell. When the adult reproduces, the hope is that the new modified DNA will pass to its young.

Transgenic Sheep

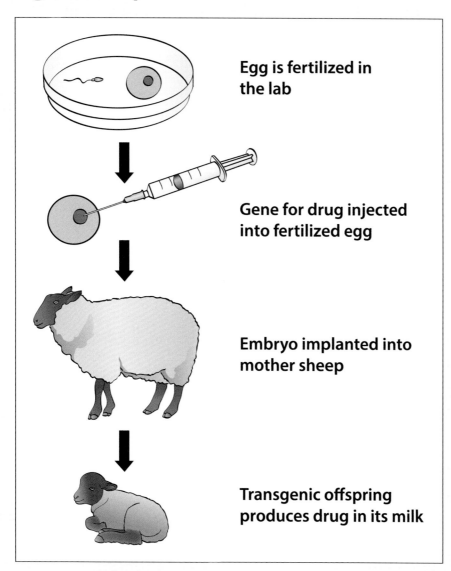

Egg is fertilized in the lab

Gene for drug injected into fertilized egg

Embryo implanted into mother sheep

Transgenic offspring produces drug in its milk

Genetic engineering of livestock begins in the laboratory. The transgenic embryo is fostered by a mother sheep. A female offspring grows into an adult that produces the desired chemical in its milk.

Gene therapy promises to cure genetic diseases such as cystic fibrosis and sickle-cell anemia. By the early 1990s, scientists could insert copies of healthy genes into the cells of people with these diseases. They do this by splicing a piece of healthy DNA into the DNA of a virus. Then they infect a patient with the virus. The virus takes over the genetic material of the infected cells. In this process, the healthy DNA becomes part of the patient's genes.

Ethical Considerations

These technologies alter natural evolutionary processes. They influence the genetic makeup of organisms to make them beneficial to humans but can also have harmful consequences. Our task for the moment is to consider the science. The question of what to do with the science is important. Spend time discussing your thoughts with your classmates and family.

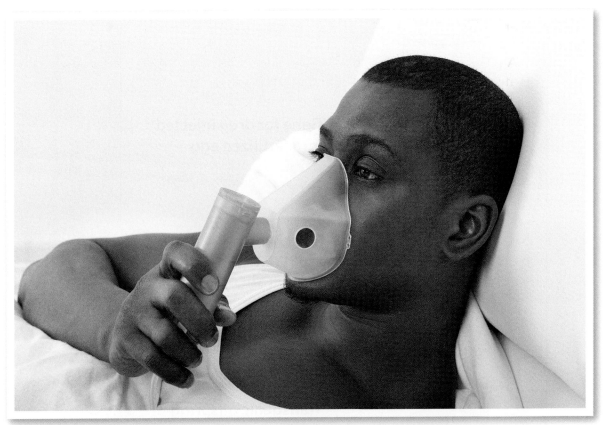

Gene therapy is an experimental technique that uses genes to treat or prevent disease. Because it is risky and still under study, gene therapy is currently being tested only for diseases that have no other cures.

Science Safety Rules

1. Always follow the safety procedures outlined by your teacher. Follow directions, and ask questions if you're unsure of what to do.

2. Never put any material in your mouth. Do not taste any material or chemical unless your teacher specifically tells you to do so.

3. Do not smell any unknown material. If your teacher asks you to smell a material, wave a hand over it to bring the scent toward your nose.

4. Avoid touching your face, mouth, ears, eyes, or nose while working with chemicals, plants, or animals. Tell your teacher if you have any allergies.

5. Always wash your hands with soap and warm water immediately after using chemicals (including common chemicals, such as salt and dyes) and handling natural materials or organisms.

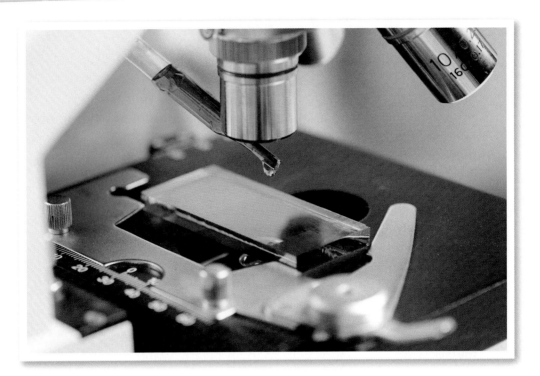

6. Do not mix unknown chemicals just to see what might happen.

7. Always wear safety goggles when working with liquids, chemicals, and sharp or pointed tools. Tell your teacher if you wear contact lenses.

8. Clean up spills immediately. Report all spills, accidents, and injuries to your teacher.

9. Treat animals with respect, caution, and consideration.

10. Never use the mirror of a microscope to reflect direct sunlight. The bright light can cause permanent eye damage.

Glossary

adaptation any trait of an organism that increases its chances of surviving and reproducing

allele variations of genes that determine traits in organisms; the two corresponding alleles on paired chromosomes constitute a gene

artificial selection selective breeding

atom the smallest particle of an element

body fossil a rock made from parts of an organism such as bones or teeth

brachiopod an early marine organism with two hard shells

characteristic a trait that helps identify an organism

chromosome a structure made of coiled DNA that transfers hereditary information to the next generation

cladogram an evolutionary tree diagram based on shared characteristics

common ancestor an organism from the past that is related to all the organisms in the group

deoxyribonucleic acid (DNA) a molecule that contains an organism's genetic information

descendant an organism related to an organism that lived earlier

dominant a form of gene that is expressed as the trait when a dominant allele is present

era a broad time span based on typical life-forms

evolution the process by which modern organisms have descended from earlier life-forms

evolve to change

F_1 generation (first filial) the offspring of the parent generation

F_2 generation (second filial) the offspring of the F_1 generation

feature a structure, characteristic, or behavior of an organism, such as eye color, plant height, or timing of migration

filial related to sons and daughters

fossil any remains, trace, or imprint of animal or plant life preserved in Earth's crust

fossil record all the fossils on Earth

gene the basic unit of heredity carried by the chromosomes; codes for proteins which determine the traits of an organism

gene therapy the process of splicing a piece of healthy DNA into the DNA of a virus so that the healthy DNA becomes part of the patient's genes, replacing the defective genes

generation offspring that are at the same stage of descent from a common ancestor

genetically modified organism (GMO) an organism whose genetic material has been altered using genetic engineering

genetics the study of genes and how they affect the traits of an organism

genome an organism's complete set of DNA

genotype an organism's particular genetic makeup

geologic time the period of time ranging from the formation of Earth about 4.6 billion years ago to today

geologist a scientist who studies Earth, its materials, and its history

heredity the passing of traits from parent to offspring

heterozygous a gene composed of two different alleles (a dominant and a recessive)

homozygous a gene composed of two identical alleles

human genome a human's complete set of DNA

inference an explanation or assumption that people make based on their knowledge, experiences, or opinions

inheritance the passing on of genetic traits from parents to offspring

inherited characteristic a genetic trait that an ancestor passes on to its descendants

isotope a variation of an element

most recent common ancestor the first organism from the past that is related to all the organisms in the group

mutation a random change in an organism's DNA; a mutation can have positive, negative, or neutral effects on the organism

natural selection the process by which the individuals best adapted to their environment tend to survive and pass their traits to subsequent generations

organism a living thing

P generation (parent) the first generation in a group of organisms that are being studied

paleontologist a scientist who studies fossils

particle a small piece of a substance that is still that substance

phenotype the traits produced by the genotype; the expression of genes

population all the individuals of one kind in a specified area at one time

principle of superposition a theory that says sedimentary rocks on the bottom are older than rocks on the top

protein a molecule that is part of cells and cell structures; determines the traits of an organism

Punnett square a mathematical model that predicts the probability of possible genotypes and the phenotypes resulting from a genetic cross

radioactive isotope an unstable isotope that decays at a predictable rate

recessive a form of a gene that is expressed only when a dominant allele is not present

related belonging to the same group or family; connected by common ancestry

sediment pieces of weathered rock such as sand, deposited by wind, water, and ice

sedimentary rock a rock that forms when layers of sediments are stuck together

species a group of organisms that can interbreed or pass on its genes to following generations

tetrapod a vertebrate with four limbs

theory a well-tested scientific explanation

theory of evolution the theory that all species are descended from common ancestors who have changed over time by means of natural selection

trace fossil a rock that preserves evidence such as animal tracks or impressions

trait the specific way a feature is expressed in an individual organism; for example, blue and green eyes are traits for eye color

transgenic organism an organism whose DNA has been altered through genetic engineering

trilobite an extinct saltwater organism that lived on the ocean floor 525–250 million years ago

variation the range of expression of a trait within a population

Each puppy in this litter received a unique combination of genes, half from its mother and half from its father. So each pup is similar, but not identical, to its parents and siblings.

Index

A
adaptation, 41–52, 54, 65, 91
Adélie penguin, 43, 44–46, 50
allele, 32–35, 91
antibiotic, 53, 58
artificial selection, 85, 91
atom, 5, 91

B
bacteria, 56
blue moon butterfly, 56–57
body fossil, 3, 15, 91
brachiopod, 10, 91
BRCA gene, 39
breast cancer, 39

C
characteristic, 17, 19, 91
chromosome, 25–27, 36, 91
Clack, Jennifer, 11–16
cladogram, 18, 19, 21, 91
common ancestor, 18, 91
Crick, Francis H., 26, 27

D
Darwin, Charles, 18, 62–67
deoxyribonucleic acid (DNA)
   defined, 20, 91
   mutation, 49
   natural selection, 54
   sequence, 37
   structure, 26–27
descendant, 18, 91
disease and human genome, 38–39
dominant, 32–35, 91

E
era, 9–14, 91
evolution
   defined, 3, 91
   relationship, 21
   theory of, 61, 65, 92
evolve, 19, 91

F
F_1 generation, 30, 31, 34, 91
F_2 generation, 31, 32, 91
feature, 28, 91
filial, 30, 91
fossil
   defined, 3, 91
   study of, 11–15, 21
fossil dating, 3, 6–7, 9–10
   relative dating, 4–5, 9
fossil record
   defined, 9, 91
   evidence, 11, 51, 66
   relationship, 15
Franklin, Rosalind, 26

G
gene
   defined, 27, 91
   disease and —, 38–39
   inheritance, 32
   Punnett square, 33–35
gene therapy, 88, 91
generation
   adaptation and reproduction, 65
   defined, 23, 91
   heredity, 28–35, 64
   natural selection, 54
genetically modified organism (GMO), 86, 91
genetic information, 21, 26, 27
genetics, 28, 33, 91
genome, 36, 91
genotype, 27, 91
geologic time, 3, 8–10, 92
geologist, 3, 9, 92

H
heredity
   chromosomes, 25
   defined, 23, 92
   deoxyribonucleic acid (DNA), 26
   described, 22–23
   genes, 27
   research, 24
heterozygous, 34, 92
homozygous, 33, 92
human genome
   defined, 36, 92
   disease and —, 38–39
   ethical considerations, 40
   mapping, 36–40
Human Genome Project, 36, 37
Hurst, Gregory, 57

I
inference, 34, 92
inheritance, 23, 24, 32, 92
inherited characteristic, 18, 92
isotope, 6, 92

J
Jarvik, Erik, 12

K
Koehler, Ken, 53, 58

L
larkey, 55
Linnaeus, Carolus, 17, 18

M
Mendel, Gregor, 24, 25, 27, 28–35, 36, 66
most recent common ancestor, 18, 20, 92
mutation
   adaptation, 49, 51, 52
   defined, 39, 92
   natural selection, 54, 58

N
natural selection, 53–59, 65–66, 92

O
offspring
   adaptation, 49
   natural selection, 54, 55
   Punnett square, 34
   trait, 35
organism, 3–4
   adaptation, 42, 48, 50, 51, 65
   classification, 17
   dating, 7
   defined, 3, 92
   eukaryotic, 36
   fossils, 15
   genes and heredity, 27
   human genome, 40
   natural selection, 54
   relationship, 21
   variation in traits, 23

P
P generation (parent generation), 29, 30, 31, 92
paleontologist, 5, 9, 12, 16, 92
paleontology, 13
particle, 4, 92
phenotype, 33, 92
population, 34, 54, 56, 92
principle of superposition, 5, 92
protein, 27, 49, 92
Punnett, Reginald, 33
Punnett square, 33–35, 92

R
radioactive isotope, 6, 92
recessive, 32–35, 92
related, 18, 21, 92
reproduction, 51, 65

S
safety rules, 89–90
Salmonella typhimurium, 53, 58
Scott, Eugenie, 67
sediment, 4, 92
sedimentary rock, 4, 9, 92
species, 18, 50, 62–63, 92
Steno, Nicolaus, 5
survival, 43, 51, 65
Sutton, Walter S., 25, 33

T
tetrapod, 11–15, 93
theory, 60–61
   confirmation and explanation, 64–65
   defined, 60, 93
   description, 68
   pattern, 62–63
   publication, 65–66
   revision, 67
theory of evolution, 61, 65, 93
trace fossil, 3, 15, 93
trait
   defined, 18, 93
   evolve, 19
   genes and heredity, 23, 27, 28, 32, 34, 35, 64, 65
   natural selection, 54
   risk for disease, 38
   species global variation, 62
   survival, 43
transgenic organism, 87, 93
tree thinking, 17
   branch pattern, 18
   cladogram, 19–21
trilobite, 10, 93

V
variation, 23, 93

W
wandering albatross, 43, 47–48
Watson, James D., 26, 27